Assiya Utzhanova

Chess and Maths

Assiya Utzhanova

Chess and Maths

Want to train mathematical mind, play chess

LAP LAMBERT Academic Publishing

Impressum / Imprint

Bibliografische Information der Deutschen Nationalbibliothek: Die Deutsche Nationalbibliothek verzeichnet diese Publikation in der Deutschen Nationalbibliografie; detaillierte bibliografische Daten sind im Internet über http://dnb.d-nb.de abrufbar.

Alle in diesem Buch genannten Marken und Produktnamen unterliegen warenzeichen-, marken- oder patentrechtlichem Schutz bzw. sind Warenzeichen oder eingetragene Warenzeichen der jeweiligen Inhaber. Die Wiedergabe von Marken, Produktnamen, Gebrauchsnamen, Handelsnamen, Warenbezeichnungen u.s.w. in diesem Werk berechtigt auch ohne besondere Kennzeichnung nicht zu der Annahme, dass solche Namen im Sinne der Warenzeichen- und Markenschutzgesetzgebung als frei zu betrachten wären und daher von jedermann benutzt werden dürften.

Bibliographic information published by the Deutsche Nationalbibliothek: The Deutsche Nationalbibliothek lists this publication in the Deutsche Nationalbibliografie; detailed bibliographic data are available in the Internet at http://dnb.d-nb.de.

Any brand names and product names mentioned in this book are subject to trademark, brand or patent protection and are trademarks or registered trademarks of their respective holders. The use of brand names, product names, common names, trade names, product descriptions etc. even without a particular marking in this works is in no way to be construed to mean that such names may be regarded as unrestricted in respect of trademark and brand protection legislation and could thus be used by anyone.

Coverbild / Cover image: www.ingimage.com

Verlag / Publisher:
LAP LAMBERT Academic Publishing
ist ein Imprint der / is a trademark of
OmniScriptum GmbH & Co. KG
Heinrich-Böcking-Str. 6-8, 66121 Saarbrücken, Deutschland / Germany
Email: info@lap-publishing.com

Herstellung: siehe letzte Seite /
Printed at: see last page
ISBN: 978-3-659-46366-2

Copyright © 2013 OmniScriptum GmbH & Co. KG
Alle Rechte vorbehalten. / All rights reserved. Saarbrücken 2013

Table of Contents

Chess and Maths 3

What has maths hidden in chess 4

Computer chess 5

Usage of maths to analyze chess 23

Strategy based on Evaluation Function 28

How they work together 35

Bibliography 47

We all perceive the game of chess from a different perspective. As a great mathematician G. H. Hardy said, "Chess is an exercise in pure mathematics". This manuscript covers mathematical links with chess.

Best Regards, Assiya

Methods:

- I assumed that when playing chess and making moves, you are actually solving a math problem, but you are not aware of it.
- In order to find out if my assumption is true I have studied all the possible combinations for a white king via minimax algorithm.

During any game of chess, knowledge of chess figures and their moves is not enough, you should also anticipate future positions and possible moves – often several moves ahead.

For example, if to look for the chess board on the right, the shortest path for white king (e1) to reach position "e8" seems 7 consistent moves ahead to e2 → e3 → e4→ e5→e6→ e7 → e8. So for the king minimum is 7. However, is the straight path of "e" line is the only alternative to reach the "e8" by minimum of seven moves? Math doesn't think so...

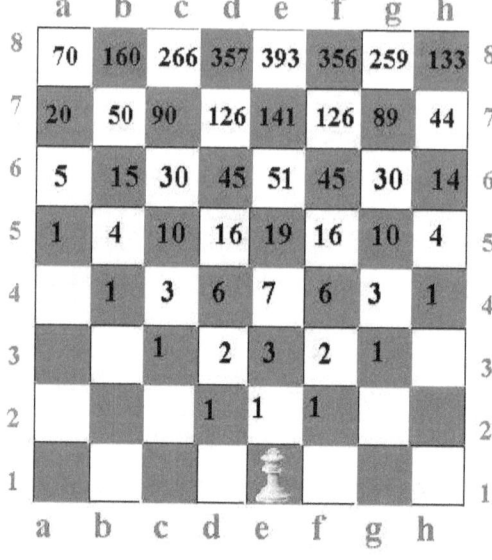

▶ If to look at the same position with white king again, you can see that all alternative moves ahead are shown by a square and presented in numbers. The picture above demonstrates a combination, when x2 is first move, which has 1+1+1=3 moves to reach d2, e2, f2 positions. Next x3, which has 1+2+3+2+1=9 moves to reach c3, d3, e3, f3, g3.

Just by counting several moves ahead and doing algebraic addition, I have come up to a solution where there are 393 possible alternatives to reach position of "e8" with 7 moves.

For example,

1) E2 → d3 → c4 → b5 → c6 → d7 → e8.
2) E2 → e3 → f4 → e5 → f6 → f7 → e8…and 391 more cases for white king.

Once I found out about a chess tournament of Garry Kasparov against computer chess Deep Blue, the fact that computer held a victory over the 13th undisputed world champion amazed me. I assumed that if the computer plays suitably good game of chess, then it uses effective computing program. This led to the bridge between mathematics and chess. This manuscript discovers links between chess and maths by using a chess computer. As it is an ideal model to start with machine programming as well as it uses mathematical language to program chess.

Deep Blue is a computer that is able to play chess. It was designed by IBM Corporation and its 6 scientists in 1997. This machine held a victory over Kasparov in a six-game match by 2 wins of computer to 1win for Garry with 3 draws. [1]

According to IBM Deep Blue research page the top 3 dissimilarities between play of Deep Blue and Garry Kasparov are as following:

1. Deep Blue evaluates and examines approximately 200 million positions on chess board per second, while Garry Kasparov can evaluate and examine maximum of 3 positions per second.
2. Deep Blue has the ambiguity on calculation ability, while Garry Kasparov with much larger amount of chess knowledge is able to do smaller amount of calculation.
3. Garry Kapsarov uses his sense of feeling and intuition while Deep Blue is just a machine that is incapable of feeling or intuition.

Consequently, two thirds of the dissimilarities are based on mathematics' topics of position analysis which is heuristic evaluation function elements, and probability based on calculation.

---

[1] Deep Blue, IBM corporation, < http://www-03.ibm.com/ibm/history/ibm100/us/en/icons/deepblue/> [Accessed on August 1st, 2013]

One of the knowledge which is learnt when playing chess is Minimax. A rule which is used in fields of decision theory, statistics, game theory in order to minimize the possible loss for a worst case (maximum loss). Simultaneously, it is able to be functioned as maximizing the minimum gain. [2]

So in a game of chess the aim is:

- to minimize the possible loss/ moves.
- Maximizing the minimum gain/ maximizing probability of reaching the best position in a minimum amount of move.

Minimize possible loss + **Maximize possible gain** = **MINIMAX**

The goal of minimax algorithm in a game of chess is to maximize the minimum gain/ probability by reaching the best position in a minimum move, which would be 7 in the case for white king. In other words, computer assigns the position with highest probability of win and minimum moves to reach it. The minimum move to reach x8 for king would be 7. Therefore more than 7 moves to reach 8th position are not accepted.

---
[2] Minimax algorithm <https://en.wikipedia.org/wiki/Minimax> [Accessed on the May 5th, 2013]

▸ Studied about computer chess programming and figured out the base of it – mathematics, through tree diagrams.

There is an algorithm for choosing to make a particular move in chess. It's like a tree.

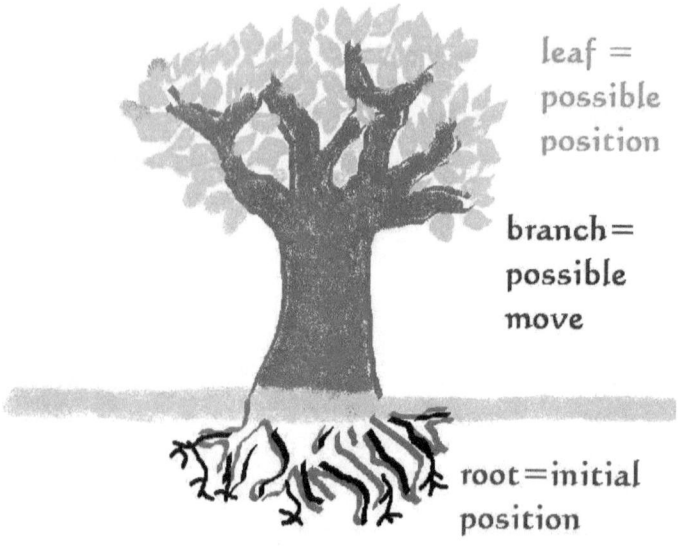

leaf = possible position

branch = possible move

root = initial position

So computer looks forward for future position from initial move, by comparison finds out the most suitable successful position ahead and makes a move of a branch leading to the following suitably successful position. This combination goes on to consecutive moves in accordance with the following algorithm.

However, some branches of the tree may lead to 1 leaf. In other words, there might be more than one alternative move in order to reach a particular position.

- Made a graph according to my investigations with white king. Human minds don't realize this, but they do these types of combinations in seconds. Computers follow the algorithm of a tree, but in graphs. They use graphs to find out the most-winning move. For example, there is a Cartesian/2 dimensions plane with x-axis and y-axis. X-axis stands for number of moves while y-axis states the moves (of a white king).

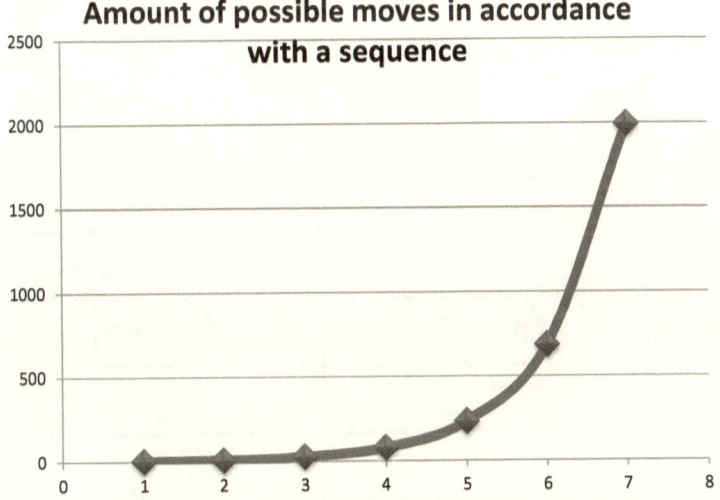

The main alternative change takes place during 4th and 6th move: 686:80= 8.6 (The graph increases 8.6 times). As you can see the hyperbole graph is a variation of a function of the inverse: $y= (k/x+b) + m$

- We shall take it for granted that sometimes moves might be inappropriate, if there's a figure on that square, or if there's a check on that position. Therefore a computer throws off all inappropriate moves which lead to those losing positions.
- In order to build the graph of the position on the board computers besides the basic knowledge of chess need mathematics.

Sequence of moves	amount of possible moves
1	3
2	9
3	27
4	80
5	235
6	686
7	1994

In the case, where there are no figures except of white king on the board, which is at the position of e1. We throw off alternative moves that take more than 7 steps in order to reach x8, which would have highest probability of win. So further we take minimum moves to reach maximum successful position by following minimax algorithm.

- Found a pattern in combinations of white king via finding probability, based on set theory.

- One of the math knowledge learned in playing chess is set theory. Chess players constantly use tree methods, Venn diagrams. The game itself is highly related to data gathering and analysis, it's relation to probability is shown on the example below. So probability represents how succesfull the following move is in percentage.

- White % represents possibility of win for white set, black % for black set, and grey % stands for a case of drawn position.

Move	Count	White %	Grey %	Black %
1.e4	597,035	38.4%	31.9%	29.7%
1.d4	455,024	38.7%	34.6%	26.7%
1.Nf3	126,616	36.8%	38.3%	24.9%
1.c4	96,598	38.5%	35.5%	26%
1.g3	11,917	38.1%	35%	26.9%
1.f4	4,378	33.7%	25.4%	40.8%
1.b3	4,183	35.8%	30.9%	33.3%
1.Nc3	1,745	32.9%	26.5%	40.6%
1.b4	1,539	37.9%	21.9%	40.2%
1.e3	492	33.9%	20.1%	45.9%
1.d3	468	33.5%	29.3%	37.2%
1.a3	301	33.9%	25.9%	40.2%
1.c3	188	31.9%	30.3%	37.8%
1.g4	133	31.6%	21.1%	47.4%
1.h3	77	26%	18.2%	55.8%
1.Nh3	22	31.8%	27.3%	40.9%
1.h4	20	50%	20%	30%
1.f3	11	54.5%		45.5%
1.a4	10	30%	60%	
1.Na3	5	60%		40%

Below you can see a tree which represents the set of chess position (the king).

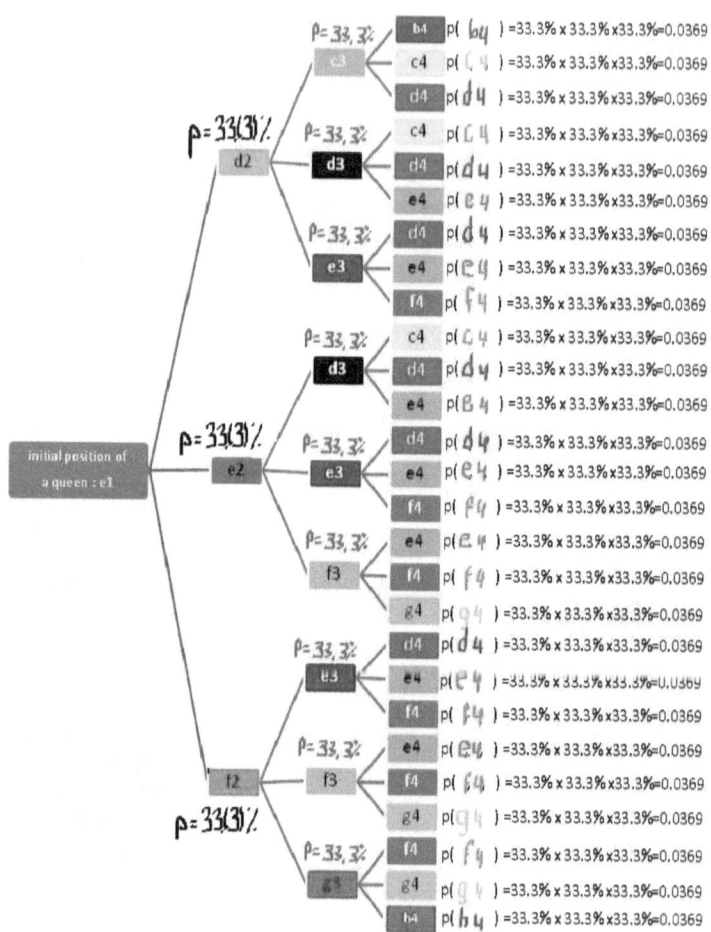

E1 is a root, initial position. X2 and X3 are the branches of tree representing second and third position moves, while X4 is the possible position.

Overall probability of possible 4th positions is as following:

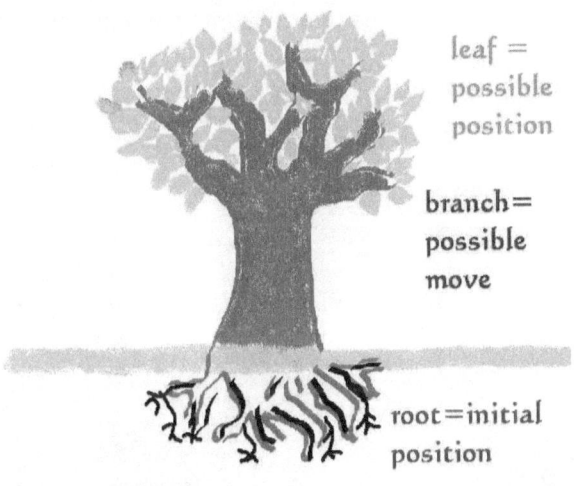

- p (b4) = 0.0369
- p (c4) = 0.0369 x 3= 0.1107
- p (d4) = 0.0369 x 6= 0.2214
- p (e4) = 0.0369 x 7= 0.2583
- p (f4) = 0.0369 x 6= 0.2214
- p (g4)= 0.0369 x 3= 0.1107
- p (h4)= 0.0369

The computer assigns a value to each ply tested. Consequently the ply with the highest probability is chosen. "Ke4" is the highest, maximum, almost 26%. Therefore it will be the most successful position for white king at the 4th position.

On the previous page minimax tree diagrams of possible 4th position were shown. Now I am taking it further...

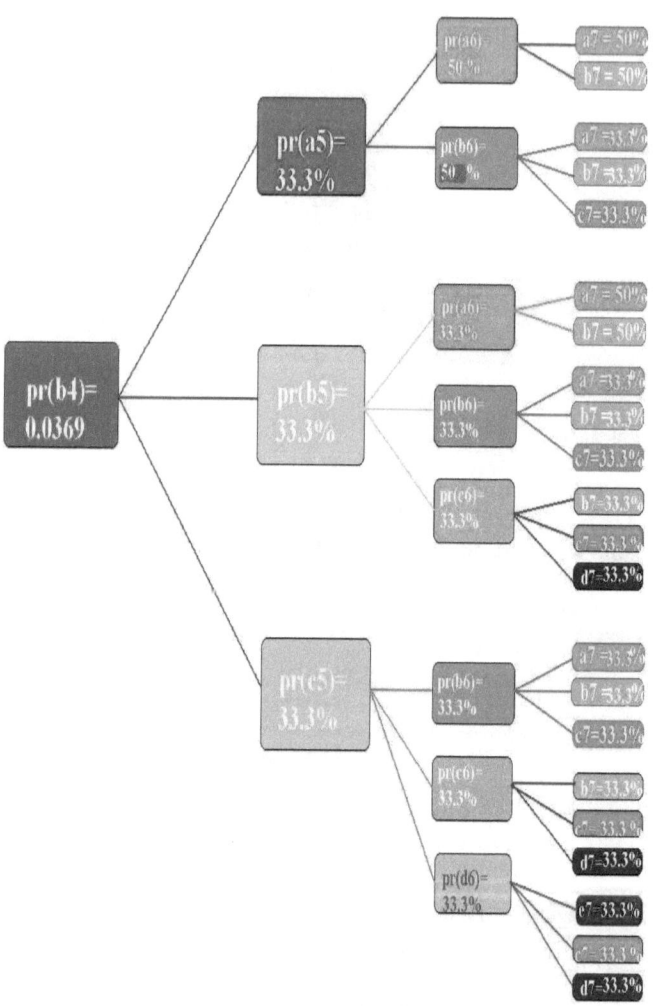

Pr(b8) =160/1994 = 0.0802

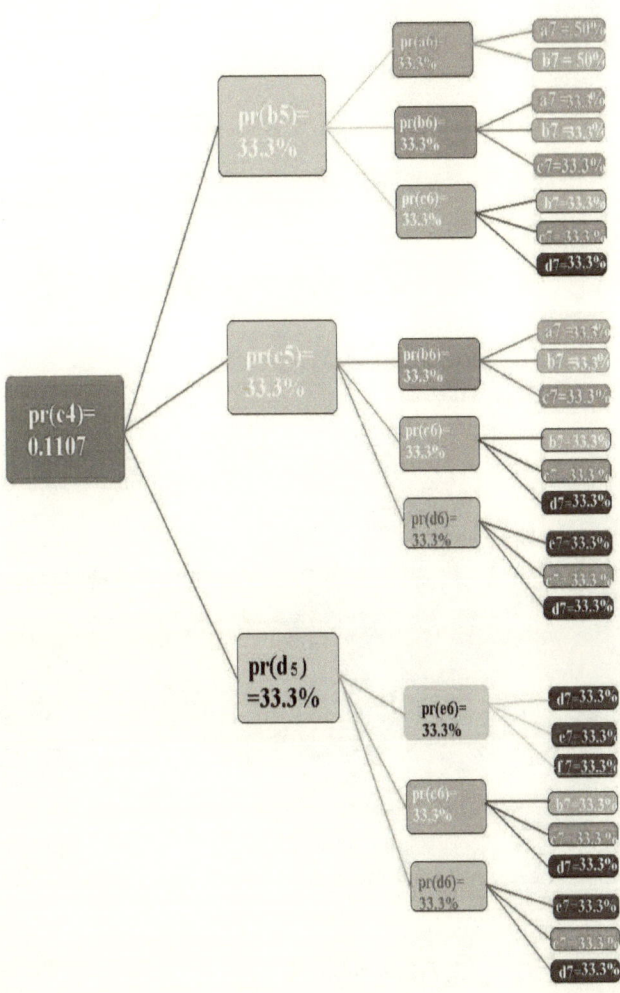

Pr(c8)= 266/1994 = 0.1334

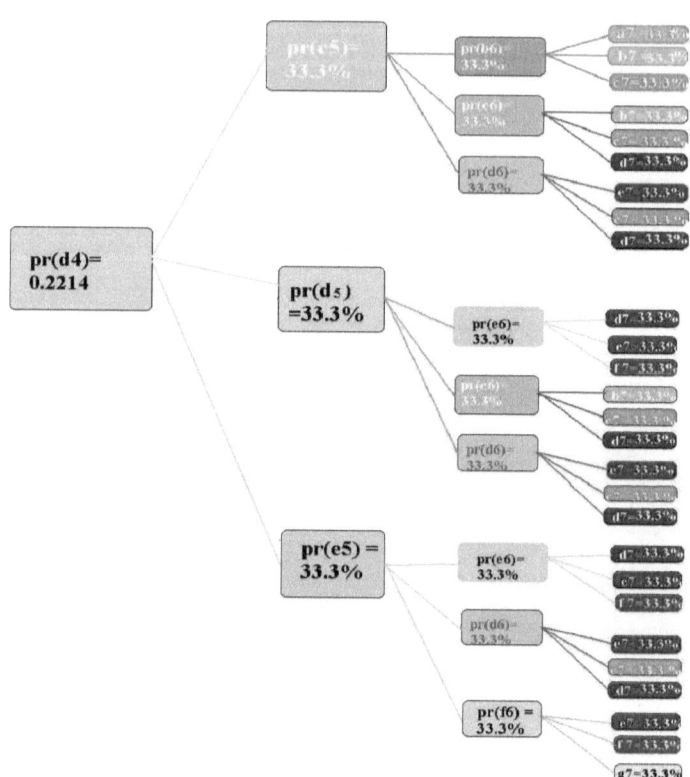

Pr(d8) = 357/1994 = 0.179

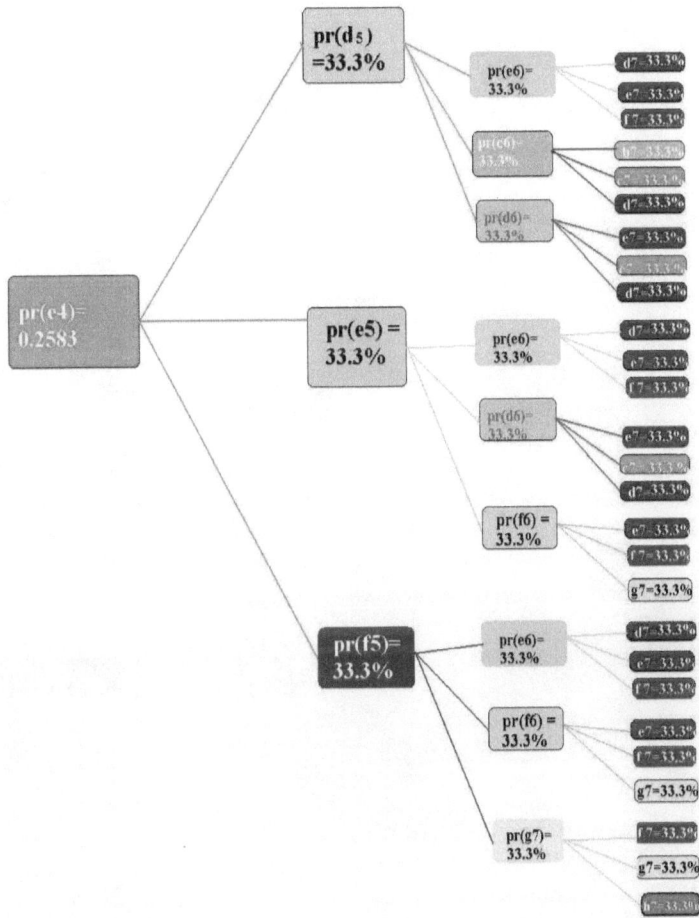

Pr(e8) = 393/1994 = 0.197

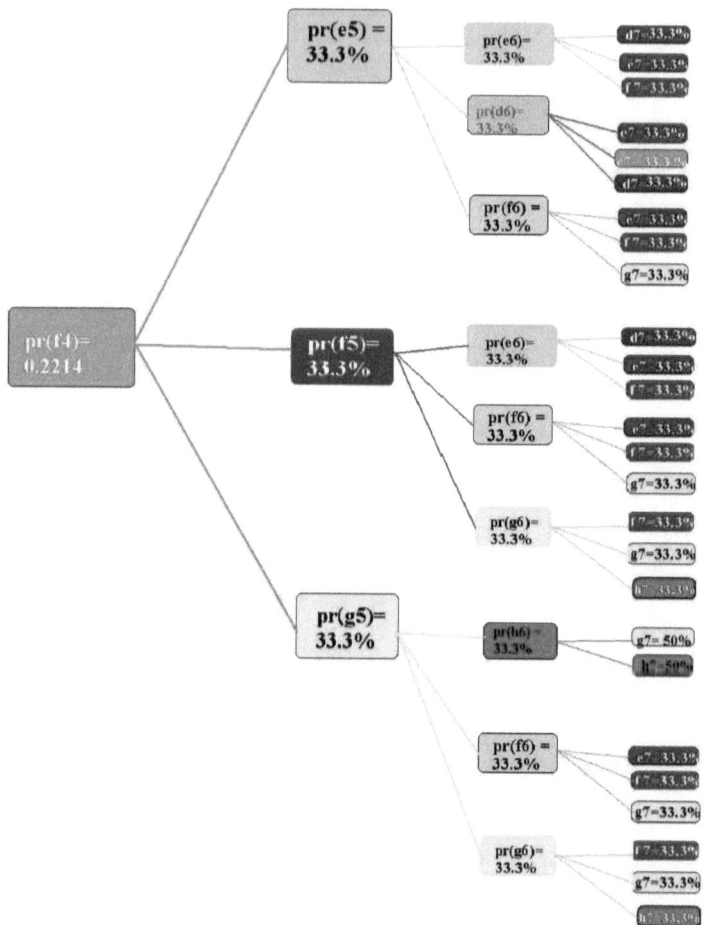

Pr(f8) = 356/1994 = 0.178

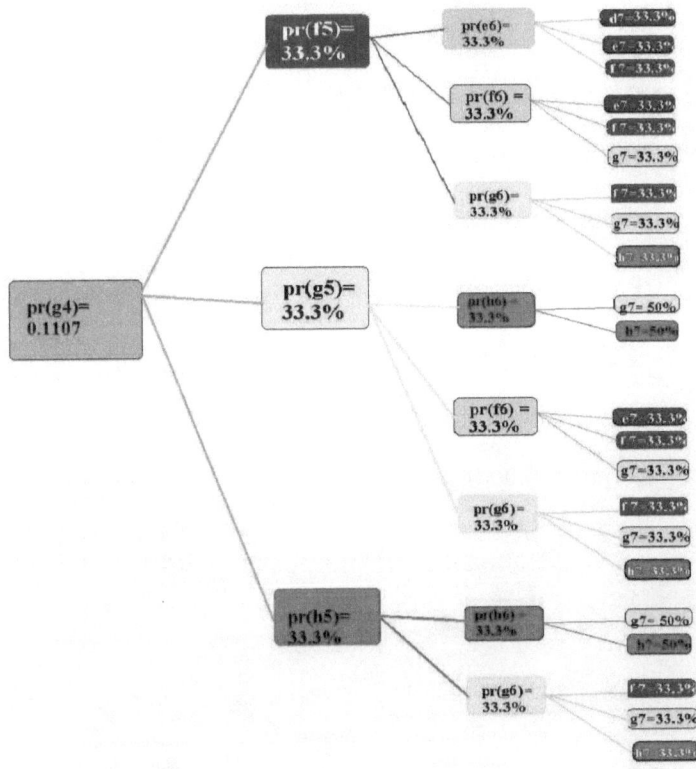

Pr(h8) = 133/1994 = 0.0667

Computer assigns the ply to the position with the best probability. Ke8 is the best alternative for the 8th position. It's probability is 19.7%

Pr(b8) =160/1994 = 0.0802

Pr(c8)= 266/1994 = 0.1334

Pr(d8) = 357/1994 = 0.179

Pr(e8) = 393/1994 = 0.197

Pr(f8) = 356/1994 = 0.178

Pr(g8) = 259/1994 = 0.1298

Pr(h8) = 133/1994 = 0.0667

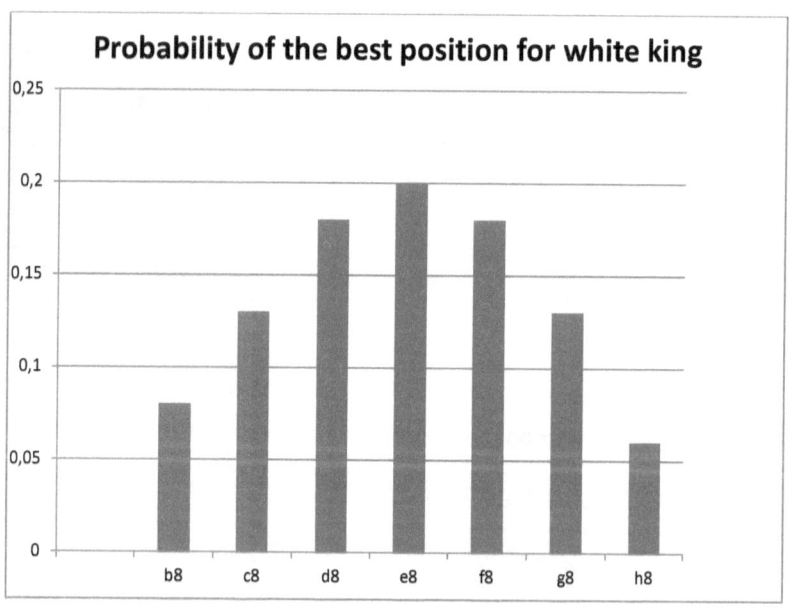

It was shown:

-how to find probability of possible positions from initial point and define the most successful position ahead via set theory and minimax algorithm .

-How to build variety of graphs according to amount of moves in a particular chess position;

-that simple algebra: calculating and logic are one of the most significant links between math and chess.

PART II

Abstract

Part II explores the most productive and efficient openings currently in a game of chess. It examines an effective strategy of consistent moves in debuts, considering the world's great masters' experience. Mathematical methods of set theory and were used to find probability of success of the openings. Heuristic Evaluation Function was used to examine productiveness of the openings.

In order to construct a hypothesis I first analyzed the powerful openings that come in manuscripts and had been written many centuries ago. They are Ruy Lopez, Sicilian Defense and Queen's gambit. I presented a list of factors that build up a powerful opening. Due to the following criteria set of 4 rules I made the hypothesis. According to my research I found out that all of them are still used currently in a game of chess. As I studied in a professional school of chess, I recorded hundreds of games. So I gathered that data and counted the moves in which White set held a victory. Then I used heuristic evaluation function conditions for a game of chess and examined the productiveness of the successful 3 moves. The hypothesis was proved.

I discovered a line where the first moves of Ruy Lopez, Sicilian Defense, and Queen's Gambit are combined. It was firstly recorded by Fischer and Auner, where the white set held a victory. According to this great tournament play of 1960's I strengthened the conclusion that my assumption is correct and the combined opening strategy of the three openings currently results productively and effectively. Because the probability of white sets wins is 50.18%. Chess evaluation strategy recorded by Claude Shannon in 1949 for computer programming led to find the links between math and chess and imagine the way computer sees the chess board, and how it thinks via numbers.

1. Introduction

"In chess so much depends on opening theory, so the champions before the last century didn't know as much as I do and other players do about opening theory. So if you just brought them back from the dead they wouldn't do well. They'd get bad openings"

–Fischer.

1.1 Topic and Focus

This part of the manuscript takes as its theme chess openings. A purpose of it is to determine the most popular and effective opening strategies nowadays via math. The main reason to study specific openings is to achieve positions in which you understand the middle-game ideas. Hence you will get a successful moves and a winning position at the end. According to Oxford Companion to Chess there are 1,327 of named openings and variations that are possible in chess.[3] Some of these variations are time-tested and very popular. Taking generally numerous of strategies are being deeply studied and wide variety of openings are being used. Based on mathematical methods I will be examining productiveness of opening strategies.

1.2 Motivation for Writing

As Emmanuel Lasker said, the game of chess survives because of many hidden and unlocked mysteries in it. Bright light of the truth would kill the interest of a chess player, who can't yet find out these treasures and is attracted by its cheeky courage and the risk of an inquirer in a search of an unknown mystery. There are many positions in the game of chess that won't be beaten in any case of a right play of partners, where you can't solve the mystery. In the simplest one you've got nothing except of 2 lonely queens on the board. There are also ones with higher probability of a victory - army of 16 white figures and pawns ready for the attack with the same amount of black. I will be discovering the mysteries of chess position like this - opening phase.

---

[3] Hooper, David; Whyld, Kenneth . *The Oxford Companion to Chess* (Oxford University Press, 1992)

1.3 The Significance of a Chess Opening

The opening face is used to develop ones pieces and control the center directly. Openings are set of the best methods of working your way into an equal or advantageous position. Understanding the basics of the moves, openings will increase your confidence, which will let you become the winner in this historically mysterious game.

There's no chess player that wouldn't show curiosity in improving their game and getting rid of disadvantages. According to the 24 lectures by Yuriy Averbuch from Academy of Chess majority people think that their only obstruction to an improvement is a non-acquaintance in a theory of debutes (openings).

1.4 Research Question

Commonly known openings come from years of trying to find the best method of play. I am facing with a question of what are the most productive openings currently in chess and how to prove it via maths.

1.5 Hypothesis

My hypothesis is that Ruy Lopez, Sicilian Defense, Queen's Gambit are the most productive and efficient openings currently in chess. I am going to find out if that is true by a play method of chess machines, which is based on mathematics.

2. Methods

"Accurate studying chess books will pave you the way

to become self-motivated and to analyze on yourself."

-Max Eyve

Young students who enter a professional school of chess in Astana on their first lesson are told these 4 lines:

"You've got a high potential and objectives,

Look for harmony and beauty of the laws.

For you honor to be heard the Olympic anthem

'Cause of reaching the richest combinations."

This saying always inspired me in the game of chess, and still inspires me now.

Every student during studying chess is taught these several rules[4]:

1. A chess game is set of 3 phases: opening, middlegame and endgame.

   I will be focusing on openings.
2. In debuts white and black sets develop their positions of figures in order to place them on profitable squares where they will show an optimal impact on the game.

   An optimal impact on the game shall be shown in the beginning, which is a part of opening.
3. You should gain the control of the center: Control of the central squares allows pieces, figures to be moved to any part of the board easily, and can show morally harmful effect on opponent.

   The control of the center is the best to be gained also in the beginning as it saves your time and gives you an opportunity to begin the attack as soon as possible.
4. King safety: It is very important to keep the king in safety; therefore you have to do the castling.

   A castling combination is when king and rook change squares. It is mostly done in the debut of a game.

   The points above demonstrate factors of a powerful opening, debut. As you can see productive opening for a game of chess should begin by development of queen's and king's pawns for gaining the control of the center; afterwards to

---

[4] S. Tarrasch, The Game of Chess(Courier Dover Publications, 1987)

deduce the queen's knight and bishop; and castling for the purpose to keep the king safe. These are the parameters I will be looking for during studying the most powerful chess openings.

Mathematics is often described as a science of pattern[5], because mathematicians commonly look for patterns and ways to express them. We find a pattern in numbers, variation, space and random events. I showed you a pattern in a game of chess. It is an example to use mathematics in different aspects of your life, and it does train mathematical mind.

Investigations showed in part 1 included only 1 player; however a game of chess requires 2. So possible chess positions are defined as,

1) statement of all positions of all pieces on the board
2) statement of white's or black's turn to move
3) statement of the king's or rook's move, because it is important for castling combination
4) Statement of the last move. It will figure out if there are possible en passant, because position's privilege might be forfeitable after 1 move.

---

[5] Tim Knight, Mathematics and Technology, Pamoja Education, 2 July 2013,
<http://news.pamojaeducation.com/mathematics-and-technology/> [August 5, 2013]

As Claude Shannon said this type of game uses evaluation function. It is applied to a position "p", which value fits one of the three categories (won, drawn, lost). It can be found by writing the number of matches in each pile in a binary notation.

WON   1

DRAWN   0

LOST   -1

The numbers are put in a column in order to make an addition. So if the number of ones in each column is even, then the position is won for the one who made the last move, and lost for the player about to move.

If to program such an evaluation function for the game the machine can play perfectly. In fact, it was programmed by Claude Shannon in 1949.

Following this evaluation function computer won't lose a position, draw a winning position, or lose a draw position. Hence it will occasionally win. If there's a mistake made by an opponent player, then computer would capitalize on it.

When it is computer's turn to move, a program calculates probability of each ply tested, various positions obtained from initial position per each possible move. It chooses a move that gives maximum value to function. Just like the investigation for white king that was shown in part I.

The relative values for each piece of a game of chess are approximately:

Queen	9
Rook	5
Bishop	3
Knight	3
Pawn	1

1) If to add the numbers of pieces of the 2 sides with given coefficients, the side with higher amount of total addition is going to have a better position.

2) Placement of rooks on open files. The side with higher mobility ends up in a winning position. Mobility stands for number of possible moves.

3) No backward, isolated, or doubled pawn as they show weakness of a position and would be highlighted with a sign of minus.

4) Exposed, unprotected and open King is weakness of the player in the opening and middle game.

According to strategy based on evaluation function pieces and their moves in a programming language that includes mathematics are expressed as following:

Computer sees it as formula

E4 – e5   Ma-Mab

Kf3 – Kc6   Mabc - Mabcd

Which leads to f(MabcdMabcMabMap)

It starts at F(P) – function of possible initial position, then computer assigns a value of "a" to the "p" of possible position, next a value of "b" of black set is assigned for the "a,p", afterwards a value of "c" of white set is assigned to "b,a,p", and a value of d is assigned for black set making computer consider the previous "c,b,a,p".

Computer takes for granted only 2 further moves. The 1st one is which it makes, and the second one is what is expected from the opponent and how to answer it. White set is always considered as a maximum and a black set as a minimum. This is important for function as it makes moves occur in a definite order. Because white have more choice as they begin and black are dependent on white.

A chess board square can be occupied by 13 different ways:

Empty : 0

White: 1-6

Black: -6-1

So the coordinate system is between -6 and 6.

Each piece's presence on the square for computer is accepted by these numbers:

Pawn (P)   1

Knight (N)   2

Bishop (B)   3

Rook (R)   4

Queen (Q)   5

King (K)   6

For example,

This is how we see the board:

This is a computer version:

```
-4 -2 -3 -5 -6 -3 -2 -4
-1 -1 -1 -1 -1 -1 -1 -1
 0  0  0  0  0  0  0  0
 0  0  0  0  0  0  0  0
 0  0  0  0  0  0  0  0
 0  0  0  0  0  0  0  0
 1  1  1  1  1  1  1  1
 4  2  3  5  6  3  2  4
```

2.1 Ruy Lopez or Spanish Strategy

Many chess players investigate and use openings of very strong world-widely-known master players. For example, Botvinnik, Karpov, Kasparov preferred to make their first move e4, which is the beginning of Ruy Lopez.

A Ruy Lopez is one of the most-developed strategies of all open games, its lines were analyzed beyond move thirty. Almost in every move the player has got many good choices, and eighty percent of them have been examined.[6] The opening was named after the 16th century Spanish priest Ruy Lopez de SEGURA that made a systematic study of the opening in a book about chess Libro del Ajedrez published in 1561.[7] Use of Ruy Lopez did not popularly develop only until the mid-19th century, because a Russian theoretician Mr. Jaenisch rediscovered its potential: the line became sharp (3...f5!?) The opening remains the most commonly used amongst the open games in master play;

It's general moves are:

1. e4 - e5

2. Nf3 - Nc6

3. Bb5 - a6

---

[6] Silman, J. "Marshall Attack" (2004)
[7] Ruy Lopez, Wikipedia The Free Encyclopedia <https://en.wikipedia.org/wiki/Ruy_Lopez> [Accessed on the 3rd of March]

```
-4  0 -3 -5 -6 -3 -2 -4
-1 -1 -1 -1 -1 -1 -1 -1
 0  0 -2  0  0  0  0  0
 0  3  0  0 -1  0  0  0
 0  0  0  0  1  0  0  0
 0  0  0  0  0  2  0  0
 1  1  1  1  0  1  1  1
 4  2  3  5  6  0  0  4
```

According to online Chess Database and Community website[8]:

The website found 1082 games in the database with the usage of C88 – Ruy Lopez opening, during the years of 1868 till 2012.

---

[8] Ruy Lopez(C88). Online Chess Database and Community, <http://www.chessgames.com/perl/chessopening?eco=c88 > [Accessed on the 17th of April]

36

The C88 opening had a huge advantage during 1890 to 1920's. Many people began using it and the peak was reached in 1910. This shows short history of usage of Ruy Lopez in a game of chess.

Statistics from the Chess Database show the overall record of:

White wins	34.8%
Black Wins	23%
Draws	42.2%

As you see it seems successful with a probability of a win in 34.8, almost 35% percent of victory in the case of you being white. To speak for a moment on my own experience, when I was nine years old I used to love the Ruy Lopez's strategy, it seemed simple and successful for me too. Going back to 2007, month of May during my participation in Kazakhstani Chess Tournament/Championship my 18 games were recorded; verses Moldagaliyev, Ashimova, Kuanishbay, Kylyshbekov, Kabdeshov, Amangeldy, Minakanova and Baizhanov. Ten of the eighteen tournament games were begun by the famous Ruy Lopez move e4-e5.

In fact I lost two of the ten games started by the Spanish opening strategy. I held a victory over 6 plays, and twice ended up with a draw. Consequently, reached a result of:

White wins	40%
Black Wins	40%
Draws	20%

It seems fair for me, as White either Black sets have the same probability of a victory.

In Ruy Lopez white and black sets develop their positions of figures in order to place them on useful squares beginning from the first move in order for figures to gain dominance over the centre. Especially, pawn, knight and bishop are in charge. The operation for king safety is regularly done within five moves, and is capable to be done before or after according to choice of the player. For example, when I played with Amangeldi, representative of chess school of Kokshetau region, a castling move took place on the seventh turn. This brief analysis clarifies productivity of the strategy.

On the other hand, opponent may response to the first move of Ruy Lopez e4 not by e5. This is another variation and another case. Let's consider one of the most common first responses by black c5, which is Sicilian Defense.

2.2 Sicilian Defense

"I'm really afraid when my opponent moves c5 after I play 1. e4" says Fevil from Spain. [9] I was curious why. After it was found out that Sicilian is a very strong and powerful opening; the black response 1...c5 proves it.[10]

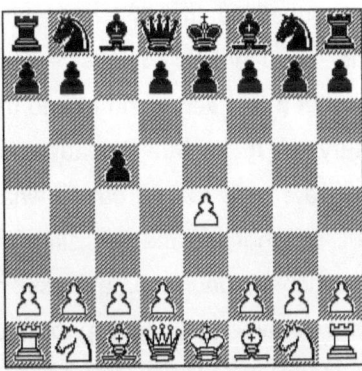

```
-4 -2 -3 -5 -5 -3 -2 -4
-1 -1  0 -1 -1 -1 -1 -1
 0  0  0  0  0  0  0  0
 0  0 -1  0  0  0  0  0
 0  0  0  0  1  0  0  0
 0  0  0  0  0  0  0  0
 1  1  1  1  0  1  1  1
 4  2  3  5  6  3  2  4
```

Because by open move e4 you get to attack, you get the initiative and space. However the c5 breaks the plan as it is spontaneous and keen. This move is followed by the mainline Sicilian, which makes the position for black winning, by taking over a control of d4, the centre.

---

[9] How do you play against Sicilian Defense? , Chess.com
<http://www.chess.com/forum/view/chess-openings/how-do-you-play-against-sicilian-defense> [Accessed on the 12th of May]
[10] Sicilian Defense, The Chess Website. <http://www.thechesswebsite.com/sicilian-defense/> [Accessed on the 1st of May]

[11]The Sicilian Defense comes from analyze of Guilio Polerio in his manuscript written in 1594. However it was not called Sicilian defense at that time. Later in 1893 English master J. Saratt standardized it from Italian phrase "il giocho siciliano" (The Sicilian Game) to Sicilian Defense. The question is why was it called Defense? Why do people confirm it as opening, when it is a defense? What is its efficiency about? Danican Pheldor has answered these questions in 1777. He said, "This way of opening the game ... is absolutely defensive, and very far from being the best ... but it is a very good one to try the strength of an adversary." It is true because of opponent's reply to 1...c5, you are able to find out to what extent can the player do well, or what are the key skills of him/her. Moreover in the opinion of Jaenisch, a Russian chess player who in 1840's was amongst the top of the World[12], and the Handbuch, German chess magazine, the defense is the best possible response to a Pawn going to e4, 'as it renders the formation of a center impracticable for White and prevents every attack.' As you can see the Sicilian Game 'produces psychological and tension factors which denote the best in modern play and gives notice of a fierce fight on the very first move'[13]. Actually that is the reason of Fevil's fear.

In Fischer's popular chess biography book *"My 60 Memorable Games"*, thirty seven percent of the games were opened by Sicilian. In fact this proves that the famous world champions

---

[11] The Sicilian Defense, Chess.com <http://www.chess.com/blog/MiramarMan/the-sicilian-defence-a-brief-history?_domain=old_blog_host&_parent=old_frontend_blog_view> [Acessed on the 25th of April]
[12] Adriano Chicco, Giorgio Porreca: Dizionario enciclopedico degli scacchi, (Mursia, Milano 1971)
[13] A brief history, Chess.com <http://www.chess.com/blog/MiramarMan/the-sicilian-defence-a-brief-history?_domain=old_blog_host&_parent=old_frontend_blog_view> [accessed on the 25th of May]

use it. However there are different perspectives about the defense and its productivness. For example, Henry Bird says, "The Sicilian ... has probably undergone more vicissitudes in regard to its estimation and appreciation than any other form of defense. In 1851, when the Great Exhibition London Tournament was commenced, it was entirely out of favor, but its successful adoption on so many occasions by Anderssen, the first prize winner, entirely restored it to confidence. Its rejection by Morphy in 1857–8, and by Steinitz in 1862, caused it again to lapse in consideration as not being a perfectly valid and reliable defense. Its fortunes have ever since continued in an unsettled state. Staunton (three weeks before his death), ... pronounced it to be quite trustworthy, and on the same date Lowenthal expressed a similar opinion. Baron Kolisch ... concurs in these views." [14]

Not every chess master player confirmed its greatness, hence it had some minuses. What are they? Why great World grandmasters as Morphy and Steinitz refused to use it in 1860's? According to a hypothesis[15] by a Canadian chess player Jimmy Vermeer, whenever Morphy started with e4 Thompson easily won by playing the Sicilian Defense. Maybe that is one of the reasons why Morphy disliked Sicilian, because of its greatness.

---

[14] Games Played in the London International Chess Tournament 1883. British Chess Magazine.

[15] P.Morphy vs J.Thompson, Online Chess Database and Community, updated Jan-03-08.
<http://www.chessgames.com/perl/chessgame?gid=1238221> [Accessed on the 12th of May]

What about Steinitz? He said, "Only once as far as I can remember, I used the Sicilian, as a second player. I always play an open game when I am on the defense, and accept any gambits that are offered, but, as first player, I have latterly adopted a safe and sound opening like the Ruy Lopez against Zukertort, and the Queen's Gambit against Tschigorin and Gunsberg, and I made up my mind not to alter the openings until I was a good number games ahead. As all those matches were pretty close I had little opportunity of varying, though in former days, when I had a clearer memory, I ventured into a varieties of an attack." Now it's time to evaluate that people were mostly afraid of the risk taken by Sicilian and tried to avoid it.

1. e4 – c5

   2... What is the next continuous move? Well, d4 is a good alternative.

It is an opening move of Queen's Gambit.

2.3 Queen's Gambit

The Queens' Gambit, also known as a Berg Defense is a chess opening that starts with following moves:

1. d4 – d5
2. c4

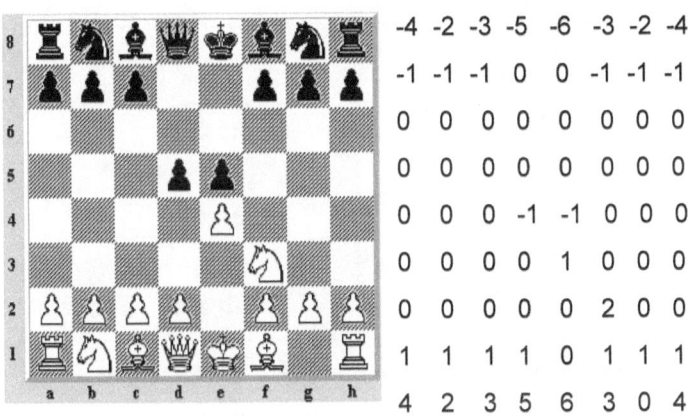

An objective of the Gambit is to sacrifice its pawn to gain dominance over a square e5.

The black set can either accept the line by eating the white pawn, or deny it by any other variation. This is one of the most popular debuts, because of daring to take a risk. White's mission is to attack and Black's mission is to defend correctly. There are some people that enjoy putting pressure on the opponent; therefore Queen's Gambit is a suitable strategy for them. It is also one of the oldest chess openings. It had been mentioned in a manuscript of 1490's and it was later analyzed. However the strategy was not popular during the early period of modern chess. Its potential was discovered during Vienna Tournament in 1973. It shows the productiveness and relevancy as it was played in 32 out of 34 games in the world championship match of 1927 among Alekhine and Capablanca. Furthermore Steinitz, the first

Undisputed World Champion developed a chess theory and increased its appreciation because of positional play. [16]

White wins	35.6%
Black Wins	45%
Draws	19.4%

According to statistics from Online Chess Database and Community Website Blacks have a higher probability of the victory, as previously White set had the ambiguity.

2.4 Fischer's interpretation of the openings

Reached three great moves by studying hundreds of openings, finding probability of win, analyzing by usage of computer-chess methods based on maths. Finally when I discovered Fischer's play verses Auner in 1960, the opening of their game was combined Ruy Lopez, Sicilian and Queen's gambit, just like in the hypothesis.

Firstly, the Ruy Lopez opening 1.e4

Secondly, The Sicilian defense key 1...c5

Thirdly, The Queen's Gambit 2. d4

---

[16] The Queen's Gambit, Wikipedia The Free Encyclopedia <http://en.wikipedia.org/wiki/Queen's_Gambit> [Accessed on the 15th of April]

Understanding that you're on the right way brings an exciting experience. These 3 together make up a strong opening strategy for White with a statistical mean[17] for:

White wins: sum of all white wins in percentage/number of possible moves = 50.18%

Black wins: sum of all black wins in percentage/number of possible moves = 32.06%

Draws: sum of all draws in percentage/number of possible moves = 17.73%

3. Conclusion

I would like to return to my research question and say that the hypothesis is correct. Ruy Lopez, Queen's Gambit and Sicilian Defense are the most productive and efficient openings currently in chess. Undisputed World Champions of Chess[18], Steinitz - Queen's Gambit, Murphy – Ruy Lopez and Fischer – Sicilian Defense proved the efficiency of the openings by making them fashionable. In fact according to the gathered data their relevancy is increasing.

In retrospect, the main goal in a game of chess is to make strong and powerful moves, where you can create an unbalanced position, in which u have a higher probability of winning. According to my research and works above, I

---
[17] Definition: a number expressing the central or typical value in a set of data, in particular the mode, median, or (most commonly) the mean, which is calculated by dividing the sum of the values in the set by their number.
[18] World Chess Championship, Wikipedia The Free Encyclopedia
<https://en.wikipedia.org/wiki/World_Chess_Championship> [Accessed on the 8th of May]

introduced you the three key moves of three extremely good openings for a powerful beginning for White: attack, defense, and attack after a defense. In the following opening the probability of white sets victory is 50.18%. Therefore it is a productive opening for white set, and we are able to conclude it by mathematics' set theory, calculation and evaluation functions.

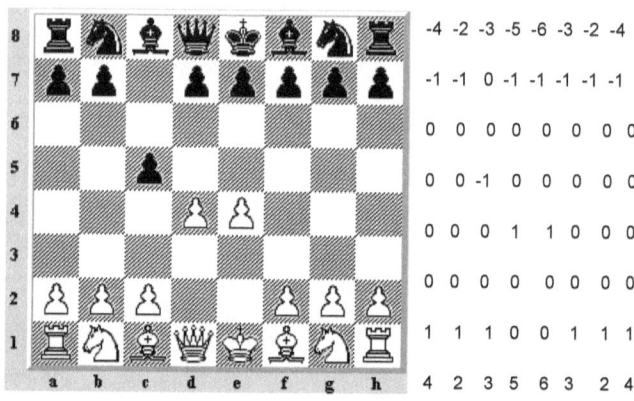

Bibliography

1. A.N. Koblenc, Chess Combinations (Moscow: 1970), page 4.
2. L. Verkhovskii, Tie (Moscow: "Physical Education and sports", 1979), page 8
3. Y. Averbach, Chess Academy (Rostov-na-Donu: "Phoenix", 2002), page 25
4. V. Ilionidov , For Beginners (Moscow: " Physical Education and Sports", 2004),
5. S. Tarrasch, The Game of Chess(Courier Dover Publications, 1987)
6. R.J. Fischer vs Auner , Online Chess Database and Community
   <http://www.chessgames.com/perl/chessgame?gid=1250832>
   [Accessed on the 5th of May]
7. J.I. Minchin (editor) (1973 (reprint)). *Games Played in the London International Chess Tournament 1883*. British Chess Magazine.
8. Oxford Dictionaries, Oxford University Press
   < http://oxforddictionaries.com/definition/english/average?q=average>
   [Accessed on the 12th of May]
9. Claude E.Shannon, Programming a Computer for Playing Chess (Philosophical magazine No. 314, 1950)

i want morebooks!

Buy your books fast and straightforward online - at one of world's fastest growing online book stores! Environmentally sound due to Print-on-Demand technologies.

Buy your books online at
www.get-morebooks.com

Kaufen Sie Ihre Bücher schnell und unkompliziert online – auf einer der am schnellsten wachsenden Buchhandelsplattformen weltweit! Dank Print-On-Demand umwelt- und ressourcenschonend produziert.

Bücher schneller online kaufen
www.morebooks.de

 VDM Verlagsservicegesellschaft mbH
Heinrich-Böcking-Str. 6-8
D - 66121 Saarbrücken

Telefon: +49 681 3720 174
Telefax: +49 681 3720 1749

info@vdm-vsg.de
www.vdm-vsg.de

www.ingramcontent.com/pod-product-compliance
Lightning Source LLC
Chambersburg PA
CBHW031550210526
45464CB00003B/1240